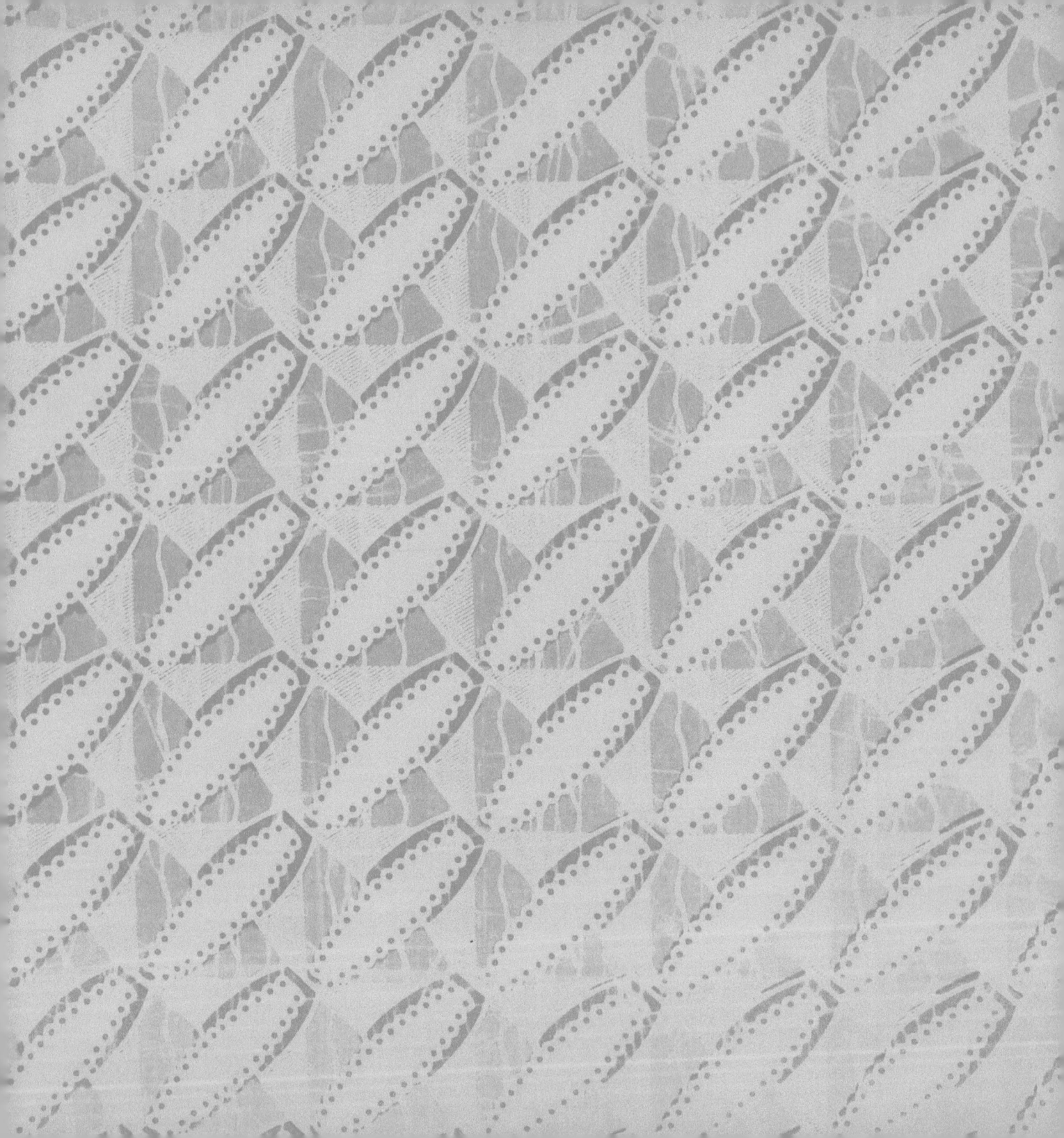

THIS BOOK BELONGS TO

....

....

....

For Zuri and Alex

The author would like to thank everyone
who helped in poducing this book,
especially Fraser Ntukula, Carey Burke
Anna Simonite and Kelvin Ntukula.

Text by ©2016 Olive Elmer Burke

Illustrations by
©2016 Fraser Ntukula
©2016 Kelvin Ntukula

ISBN 978-1532704062

@oliveelmer

www.oliveelmerburke.com

When Neema went on a journey

A happy counting adventure written by Olive Elmer Burke

Illustrated by Fraser Ntukula and Kelvin Ntukula

Neema went on a journey one day,
And saw a few things coming her way.

One boy with a stick
hopping so delicately,

So Neema hippity
hopped fearlessly.

Two cows on a farm
looking so seriously,

So Neema moo moo'd
at them playfully.

Three masks on a stall
hanging so neatly,

So Neema counted them
all up confidently.

Four drums on the ground
being played so loudly,

So Neema wiggle
wobble wiggled joyfully.

5

Five women at the borehole chatting casually,

So Neema waved back cheerfully.

Six pots soaking in
water so bubbly,

So Neema popped
the bubbles proudly.

Seven mangoes so fresh
laying there so boldly,

So Neema picked
them up bravely.

Eight jumping fish
swimming so gracefully,

So Neema stompty
stomped beautifully.

Nine birds in a tree
sounding so chirpily,

So Neema listened
on carefully.

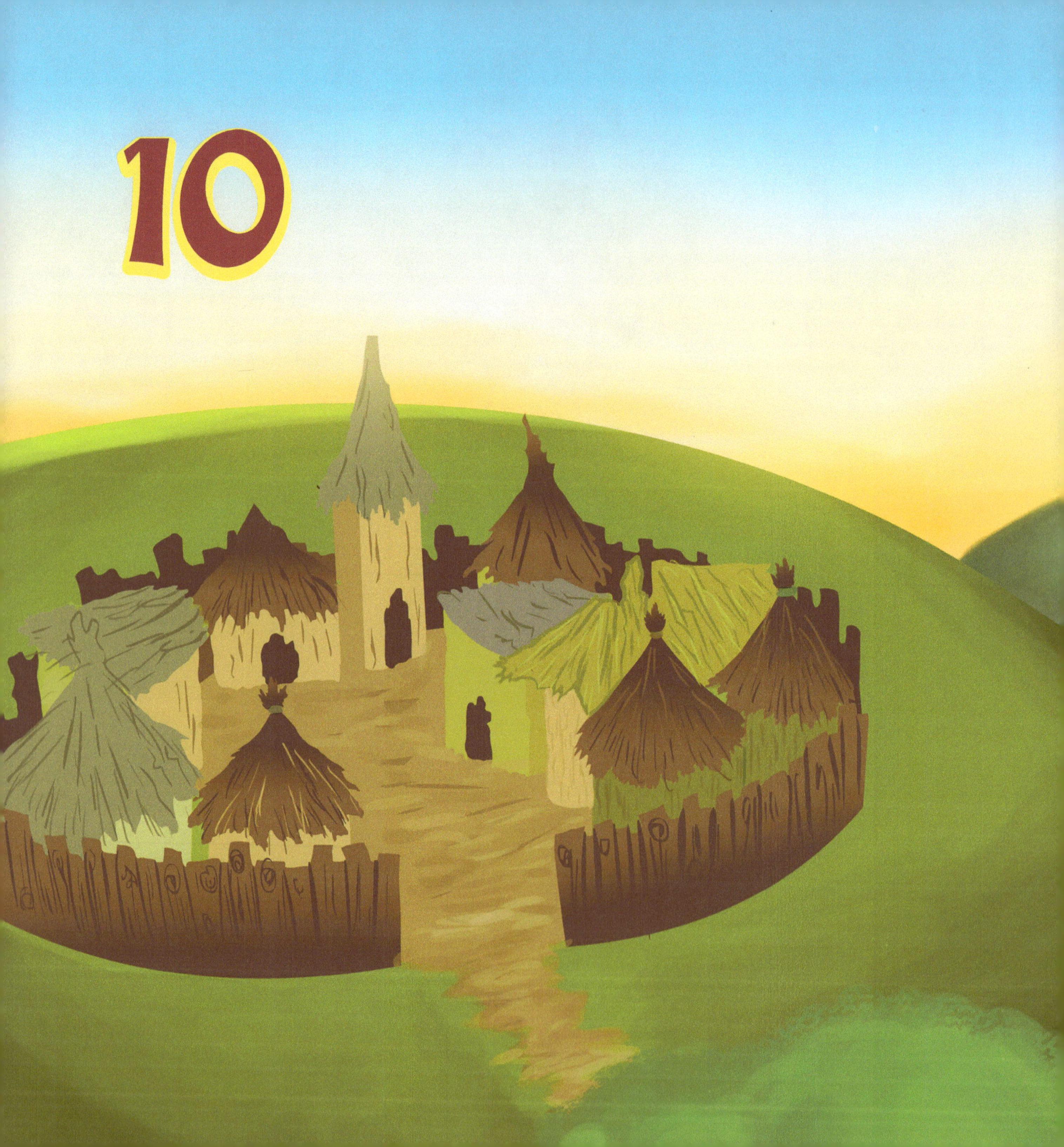
10

Ten huts down the hill
standing so firmly,

So Neema yelled out
yoo- hoo! ever so proudly.

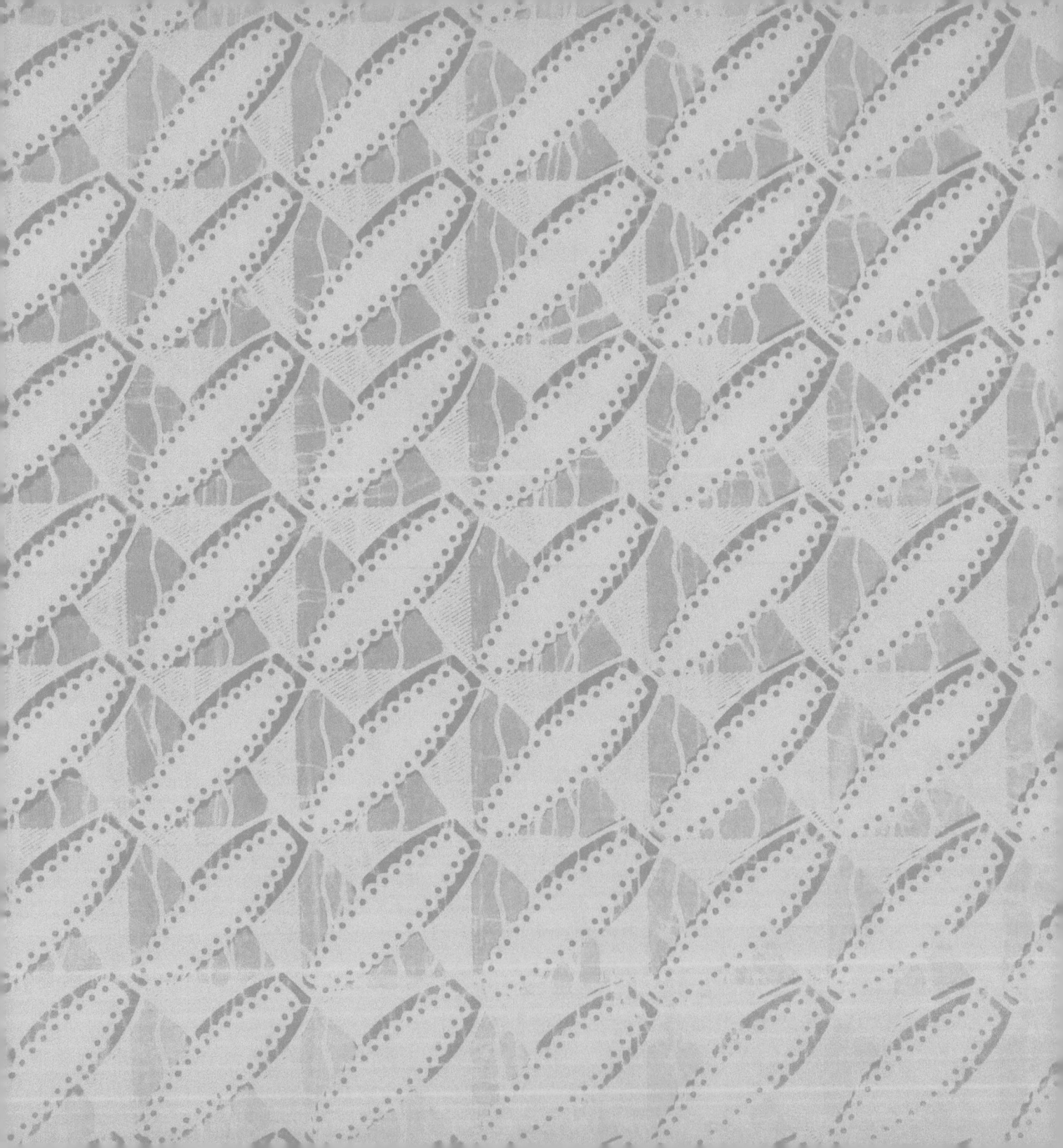